T0408171

Published by
Princeton Architectural Press
A division of Chronicle Books LLC
70 West 36th Street
New York, NY 10018
papress.com

Originally published as
Tentacoli: Piccolo Catalogo di Polpi, Seppie e Calamari
by Nomos Edizioni
via Piave, 15
21052 Busto Arsizio (VA)
Italy
t +39 0331.382339
www.nomosedizioni.it

Printed and bound in China
28 27 26 25 4 3 2 1 First edition

For Nomos
Text: Marco Colombo and
Francesco Tomasinelli
Illustrations: Giulia De Amicis
Scientific Review: Andrea Impera
Design: Andrea Amato/Tipiblu.com

For Princeton Architectural Press
Translation from the Italian: Sylvia Notini
Typesetting: PA Press

Library of Congress Cataloging-in-Publication Data
Names: Colombo, Marco, author | Tomasinelli, Francesco, 1971- author | De Amicis, Giulia, 1986- illustrator
Title: Sea wonders : the octopus, the cuttlefish, and the squid / Marco Colombo, Francesco Tomasinelli ; illustrations by Giulia De Amicis.
Other titles: Tentacoli. English
Description: New York : Princeton Architectural Press, [2025] | "Originally published as Tentacoli: Piccolo Catalogo di Polpi, Seppie e Calamari by Nomos Edizioni via Piave, 15 21052 Busto Arsizio (VA) Italy."—Title page verso. | Includes bibliographic references. | Summary: "A beautifully illustrated guide to thirty species of octopuses, cuttlefish, and squids"—Provided by publisher.
Identifiers: LCCN 2024046353 | ISBN 9781797233154 ebook | ISBN 9781797233147 hardcover
Subjects: LCSH: Cuttlefish | Squids | Octopuses
Classification: LCC QL430.2 .C65 2025 | DDC 594/.5—dc23/eng/20241121
LC record available at https://lccn.loc.gov/2024046353

SEA WONDERS

THE OCTOPUS, THE CUTTLEFISH, AND THE SQUID

Marco Colombo
Francesco Tomasinelli

ILLUSTRATIONS BY
Giulia De Amicis

PA PRESS

PRINCETON ARCHITECTURAL PRESS · NEW YORK

CONTENTS

FOREWORD

I have been challenged enormously by both octopuses and cuttlefish. They've challenged me in the one thing I love most—underwater tracking. These animals have forced me to up my game, to push myself in understanding the tremendous complexity and wonder of their lives. Their teaching has allowed me to get under the skin of wild nature. In many ways they were the catalyst in my understanding of many of the parts of the Great African Seaforest, the underwater forest of giant bamboo kelp on my doorstep.

An octopus alone in the seaforest has over fifty prey species and many predators. So in just understanding her, you need to understand sixty other animals. These cephalopods, especially the tuberculate cuttlefish, probably the world's greatest camouflage artist, force me to become much better at the art of tracking. These animals are my greatest teachers in understanding nature. They are my kin, my wild family, and they taught me to see how deeply precious wild nature is—probably the most precious thing we will ever come across. Biodiversity is the life support system of our planet.

So a book like this, by Marco Colombo, Francesco Tomasinelli, and Giulia De Amicis, that documents and promotes the magnificent cephalopods in an artful, dedicated, biological, and magical way is a wonderful gift to the world, and I am honored to support it.

Craig Foster

NATURALIST AND FILMMAKER

PRODUCER, *My Octopus Teacher*

INTRODUCTION

Imagine landing on an alien planet covered with water: its inhabitants have soft bodies and long arms and can change shape and color, imitate other species, and use jet-propulsion to swim. This might sound like the beginning of a science fiction novel, but the fact of the matter is that this planet already exists, and it is much closer than you would imagine. It is the ocean, and its most surprising inhabitants, the cephalopods, have lots of tricks to teach us.

Octopuses, cuttlefish, and squids are versatile, enterprising animals, and indisputably intelligent, too. They know how to solve problems and use objects. They can escape from aquariums and recognize people—they can even feel dislike. Some species, like the flamboyant cuttlefish, are among the most multicolored creatures of the oceans, while others, like the blue-ringed octopus, are famous for their highly toxic bites. Cephalopods also include the largest invertebrates in the world: there is even a giant squid the size of a bus, though we still know very little about it. The depths of the ocean hide still more species with surprising features: transparency that allows them to drift past predators unnoticed, luminous organs that light up in the dark, doting parental care lasting for years. Today we can start to tell the story of these animals thanks to centuries of research that scientists have conducted about the sea, a world whose dynamics are still so little known to us.

Represented since ancient times in splendid mosaics and shiny coins, these marine animals have earned a place in the collective imagination. This is thanks in part to the legendary figure of the Kraken, a terrifying squid (or octopus) that was said to be capable of effortlessly dragging even the biggest galleons beneath the waves.

Today we have an ambivalent relationship with cephalopods. On the one hand, we eat them, bringing some species to the brink of extinction and devastating their habitats; on the other hand, we admire them, making them the focus of news stories, documentaries, and Oscar-nominated movies.

And yet, as you leaf through this book, and see the various species brought to life by the beautiful illustrations of Giulia De Amicis, you will realize just how valuable cephalopods are, not only from a naturalist's point of view, but aesthetically and symbolically, too. These creatures can be tiny or huge. Many are elegant and visually stunning. They are graced with alien-like features, and yet they are familiar to us.

Octopuses, cuttlefish, and squids, after all, are invertebrates that force us to reflect on the way we relate to our planet: their unexpected cognitive skills imply that mammals and birds are not the only animals with sophisticated behavior in some ways similar to ours.

Do they suffer? Do they dream? Do they feel emotions?

These are just some of the questions that this book seeks to answer: it is a personal tribute to cephalopods, the masters of survival, the cleverest quick-change artists in the oceans.

Marco Colombo, Francesco Tomasinelli

NATURALISTS, PHOTOGRAPHERS, AND SCIENCE WRITERS

THE QUICK-CHANGE ARTISTS OF THE OCEANS: OCTOPUSES, CUTTLEFISH, AND SQUIDS

MORE THAN JUST A NAME

—

We easily recognize an octopus by its eight arms connected directly to its head. Its body is shaped like a bag and is technically known as a mantle. This is the main feature of the class of mollusks that these animals belong to, known as cephalopods. The name comes from ancient Greek and it means "head foot," in the sense that their feet are attached to their heads. Because an octopus body has no bones and is very flexible (it can be distorted and is rich in muscle fasciae), and because octopuses have good vision connected to well-developed nervous systems, they can do almost anything: catch prey, walk along the ocean floor, use jet propulsion to swim, and employ adaptive camouflage better than any other creature.

The great class of cephalopods includes not just octopuses but also cuttlefish and squids, which are excellent swimmers with streamlined bodies. In addition to the standard eight arms, each squid or cuttlefish has two additional tentacles.

This class includes the biggest marine invertebrates, like the giant squid (*Architeuthis dux*), which is longer than a truck. The mantle of a cuttlefish or squid hides a semirigid structure inside, known as the cuttlefish bone or gladius, respectively. This is the vestige of a shell. Its dimensions can vary, but no matter the size, it helps maintain the animal's shape and also helps it to swim. Nautiluses, similar to the ancient ammonites that populated the seas even before the dinosaur age, are a special subgroup. A nautilus has a large outer spiral shell and as many as ninety tentacles without suckers.

Cephalopods have not always been as they are today: the oldest fossils of their ancestors resemble typical gastropod mollusks and are similar to snails, with pointy spiral shells.

MULTIUSE ARMS

—

Cephalopod arms, much like the human tongue, function thanks to a series of parallel muscle fibers that allow them to move in all directions. While some fibers pull, others remain rigid, serving as support for movement. Each arm can have hundreds of suckers (and in some species even claws). An inner cup is surrounded by an outer ring. When that ring presses against a surface—say, the skin outside of its prey—it creates a vacuum and causes the sucker to adhere to that surface. These details are visible even in ancient representations of these animals, such as those found on refined terra-cotta vases from Greece and mosaics from Pompeii. Clearly the artists who depicted these creatures possessed profound knowledge of their anatomy and felt an intimate connection to them.

In many species, including octopuses, the suckers are independent of each other and react automatically: as soon as they touch something they attach themselves to it. So why don't these animals ever get stuck to themselves? It was recently discovered that they have an internal system that recognizes skin so that suckers won't attach themselves to the body of their owner. At the same time, an octopus can cling on to another octopus, because an octopus's brain can override this control mechanism.

Suckers also transmit information on the chemistry and consistency of the objects with which they come into contact. Practically speaking, this means an octopus can "taste" something just by touching it.

Controlling eight or ten appendices and hundreds of independent suckers that act like sensors might seem an impossible task for the cephalopod's nervous system. But these mollusks have honed an

ingenious way of managing all this information: two-thirds of their neurons (nervous cells) are located around the body and not in the head, which hosts the "real" brain. This structure resembles a star-shaped network of computers that are constantly sending each other information. The outer ones manage many of the operations independently. They speed up the function of the whole, but they can also receive instructions from the central processing unit, the real brains of the operation. This is why tentacles can continue to move and react to stimuli even when they have been amputated—an excellent trick to keep potential predators busy. As if that were not enough, injured limbs can grow back, though the process is very slow and time-consuming.

If things get really bad, cephalopods can swim fast: by blasting water out of a siphon, a cephalopod can thrust itself in the opposite direction. Squids are the champions of this technique and move through water as quickly as fish, thanks in part to their hydrodynamic shape. Some species are even capable of leaping out of the water and cruising for dozens of yards by using fins that are similar to wings and arms that, once they are in the air, can spread wide to create a support surface.

Cephalopods also have ink sacs that can be emptied through those same siphons. Doing so creates a smokescreen that provides cover during an escape. The ink forms a shape that serves as a decoy, distracting the attacker while the cephalopod makes its getaway. In many species the ink contains chemical substances that diminish a predator's senses of sight and smell.

USING SKIN TO COMMUNICATE

—

Although the word "camouflage" brings the chameleon to mind, the real masters of this technique are cephalopods. The ability of these animals—especially octopuses and cuttlefish—to camouflage themselves is astonishing, and largely due to their sophisticated use of special cells called chromatophores. By controlling the size of these cells, they can vary their color. And by making them smaller, an octopus makes a cell less visible and can change the color of its skin more quickly: it can go from light gray to dark brown in a fraction of a second. As if that were not enough, the octopus can wrinkle its skin, generating a series of stripes and folds that make it look like algae. No other animal in the world is capable of doing anything of the kind so quickly and effectively, going from smooth to wrinkly in an instant.

Recent research has found that chromatophores can self-regulate based on the light level of the surrounding environment to blend into the background even more effectively. The animals can also change color deliberately: a frightened subject tends to become very light in color, while an irritated one who is ready to fight turns reddish.

It seems that octopuses even change color while resting, depending on their dreams. Cephalopods have good eyesight, and their big eyes can move independently of each other. This helps them communicate with other members of their species. Species that live deep in the ocean manage to pick up even the weakest light so they can move about and hunt in the depths.

Furthermore, some bioluminescent squids perform a chemical reaction induced by bacteria that live in symbiosis with them in specialized cells (photophores) to produce light in the darkness to attract prey, to communicate, and to fool their enemies.

NO ORDINARY INTELLIGENCE

—

Though it's ultimately wrong to compare animal intelligence, we can say with some confidence that octopuses are among the smartest of sea animals. They certainly stand out among marine invertebrates due to their ability to recognize shapes and to learn simple tasks like touching an object to earn a reward. They can be trained to wait to eat if they expect they will get twice as much food later. Some octopuses have learned to open screw-top jars to fish out crabs inside! Though the matter is still up for debate, some studies suggest they can learn by watching other octopuses—behavior that has been observed in only a few mammals until now. When housed in an aquarium, they show interest in playing, and each has its own personality, which includes liking some of the people who take care of them and disliking others. They can also identify people after not having seen them for months. Older animals are typically more cautious than younger ones, probably because they remember their past experiences, another unusual feature.

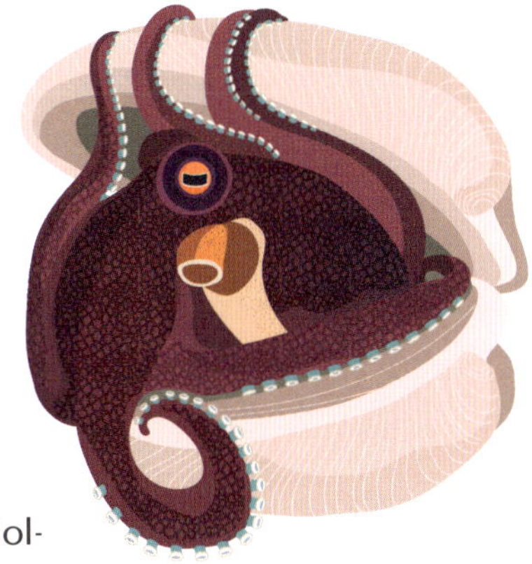

During the UEFA Euro 2008 and the 2010 FIFA World Cup, an octopus bred in an aquarium at a marine biology study center in Oberhausen, Germany, became incredibly popular because it could "predict" many of the results of the games played by the German soccer team. Paul—that was the octopus's name—got the results right eleven times out of thirteen (an 85 percent success rate) by choosing containers for food labeled with the flags of the two teams competing. Statues were dedicated to Paul the octopus and streets were named after him, even though researchers agree that the animal's predictions were likely by chance. (Besides being very lucky, this octopus seemed to prefer flags with horizontal stripes, like those of Germany and Spain. These did turn out to be the nations that played in the final.)

The octopus (which, interestingly, has three hearts and blue blood) is also among the very few invertebrates to use objects: a recent study has shown that the females of one species throw rocks at males that pester them when they don't feel like interacting.

We don't know why these animals, especially octopuses, have such sophisticated skills. Typically, these types of skills are observed in social vertebrates that live a long time and have few offspring and care for them intensely—think wolves, chimps, and dolphins. Octopuses, on the contrary, have very brief lives and tend to be solitary, except during their short mating periods. What do they do with all this intelligence, if we can call it that? The most reliable theory is that the loss of a shell (that fossils tell us was once present) led these animals to fall under intense pressure from predators. In practical terms, only the octopuses that were good at hiding or solving problems managed to survive in such a competitive environment, resulting in the amazing range of survival strategies that we see today. Scientists are very interested in this, because cephalopods developed intelligence completely independently from mammals: our common ancestor was a sea animal that lived over 600 million years ago. And yet, an octopus has around 500 million neurons, not unlike a dog.

OPPORTUNISTIC HUNTERS

—

All cephalopods are predators, capable of capturing a large variety of animals. Their limbs are their key hunting weapons: an octopus usually hooks a victim with one or two arms, and once it has established a grip, it quickly moves its body forward and covers its prey entirely before it has time to react. A cuttlefish or a squid uses its two longer tentacles for this purpose, shooting them out like arrows and then wrapping its prey in a fatal embrace.

As soon as a cephalopod has caught its prey, it uses its secret weapon—a beak similar to that of a parrot—to attack. Some of them keep prey subdued by releasing a neurotoxin carried in their saliva.

There's a little bit of everything on the cephalopod's menu. Squids catch fish and mollusks in open water, while cuttlefish and octopuses tend to prefer crustaceans found down in the depths. Octopuses also eat well-protected bivalve creatures, like scallops. These can't escape or actively defend themselves, but opening their shells is an arduous task. An octopus works hard, using its suckers and wielding its beak like a can opener. If a scallop shell proves too difficult to pry open, an octopus may pierce it and kill the scallop with poison to eat it.

When other resources are lacking, cephalopods can become cannibals, feast on unusual prey (like small sharks, seagulls, and young sea turtles), or adopt incredible strategies (one Australian species, the *Abdopus aculeatus*, actually emerges from the water to capture crabs hidden in the rocks).

In short, cephalopods are opportunists that take every chance they get, but they can also be hunted by other marine creatures, from fish to mammals. When this happens, their camouflage skills, jet-propulsion swimming, ink clouds, and intelligence raise their chances for survival.

ANYTHING FOR THEIR HATCHLINGS

—

Most cephalopods are loners and spend their lives on their own. In general, an encounter between the two sexes takes place during the mating period, at times in large groups, as is the case for many squids. With some exceptions, almost all the species stay together just long enough for the male and the female to reproduce. They do so using an arm (hectocotylus) that is shaped for this purpose. Then the female does the rest by herself: she moves away and nurtures thousands of eggs growing inside her mantle before laying them. The female octopus

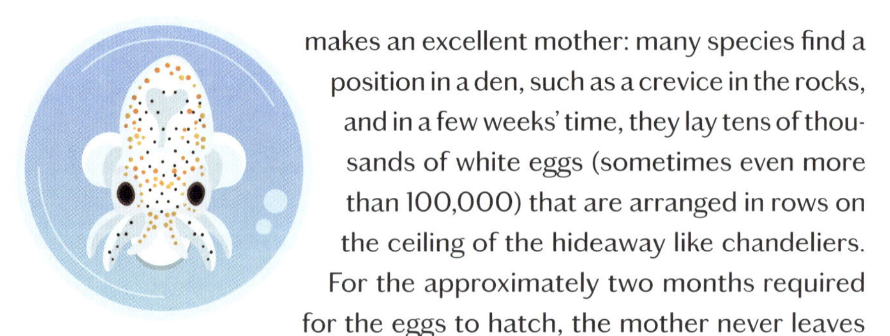

makes an excellent mother: many species find a position in a den, such as a crevice in the rocks, and in a few weeks' time, they lay tens of thousands of white eggs (sometimes even more than 100,000) that are arranged in rows on the ceiling of the hideaway like chandeliers. For the approximately two months required for the eggs to hatch, the mother never leaves the shelter, protecting the eggs from predators and using her siphon to blow water over them. Soon after the birth of the hatchlings, the mother dies.

Hatchlings are tiny larvae, less than an eighth of an inch [2 to 3 mm] long, which are immediately transported by the currents. They spend one or two months floating in plankton, but almost all of them die, eaten by small fish, jellyfish, and cetaceans. Only when they reach a few inches in length, including their arms, do they adopt an adult lifestyle. There are some species, however, that skip the planktonic larval phase completely, emerging from their eggs like miniature copies of their parents. In either case, cephalopods grow quickly: even large squids reach their full size in three years. But the lifespan of these animals is brief: most have a life cycle that ends after only two years. The reason for this lies in the way these animals react to their enemies: if every day you are at high risk of being eaten, it is unlikely that you will reach the next reproductive season alive. So the best thing to do is to invest all your energy in just one season. There are exceptions, however, such as the nautilus (which lives at

least twenty years but is protected by a shell) and the deep-sea warty octopus (*Graneledone boreopacifica*), which takes care of its eggs for four years but lives in the depths of the ocean, which means that its metabolism is slow and it has few predators.

A COMPLICATED RELATIONSHIP

—

Our relationship with cephalopods is ambivalent: on the one hand, we admire them; we dedicate movies, books, and documentaries to them; we study their arms for bioengineering purposes; and we dive deep into the water to observe them. The Netflix movie *My Octopus Teacher*, produced by the naturalist and documentary filmmaker Craig Foster of the Sea Change Project foundation, is about the day-to-day relationship between a free diver and a female octopus in South African waters. The film won the 2021 Oscar for Best Documentary Feature. On the other hand, we cook and eat these animals without giving it a second thought.

Since ancient times, octopuses, cuttlefish, and squids have aroused the interest of humans and have also been at the heart of some amazing myths and legends, such as that of the monstrous Kraken, which could drag ancient ships beneath the waves. Today, researchers conduct in-depth studies on these animals. In addition to learning about their intelligence, researchers aim to understand the lifestyle of species that live in the depths. They employ bathyscaphes and remotely operated underwater vehicles (ROVs) to observe them and use bait to attract them. Humans have been able to attach transmitters to the bodies of some, such as nautiluses, and then monitor their movements using modern technology.

Although most of the species are not in imminent danger, experts have sounded the alarm: overfishing, especially trawling, strikes animals without distinction, including those that are undersized because they have not reached maturity. Some species in the genus *Opisthoteuthis* (flapjack octopuses) are at risk, and it has been estimated that fewer than one thousand individuals remain of *Cirroctopus hochbergi* (the four-blotched umbrella octopus), which lives in the deep waters around New Zealand and is already extinct in areas where it was once abundant. Thousands of people practicing so-called recreational fishing in shallow waters by hand or harpoon, especially during the summer, affect both young and adult cephalopods. Mineral extraction on the ocean floor and the gathering of shells for collecting are considered critical for some cephalopods, like nautiluses, which are caught in large traps and then sold to tourists as souvenirs.

Currently, plans are being made to breed octopuses commercially: but to breed one 2.2-lb. [1-kg] octopus, you need 6.6 lbs. [3 kg] of fish-based feed, which means this practice is unsustainable if we are to conserve our oceans. Moreover, packing together hundreds or thousands of territorial individuals leads to a huge amount of stress and constant clashes, including cannibalism. Given that octopuses can feel both physical and emotional pain, plans like these are definitely not the way to go.

THE CEPHALOPODS

A SHELL FROM THE PAST

Although nautilus fossils are very common, few nautiluses survived extinction, and they now live in the Indian and Pacific Oceans.

A nautilus uses its shell to regulate its position in the water by means of chambers that are connected by a thin siphon. The shell is sturdy and resembles a snail shell with a large opening turned upward. This is where the nautilus's head comes out, with more than ninety thin tentacles and big eyes that can see in the dark at the bottom of the ocean.

The most famous species is the emperor nautilus (*Nautilus pompilius*), which, unfortunately, is often sold as a souvenir. These shells are widely available in stores, but don't buy them! They are fished using chicken feet as bait. Other nautiluses are rarer still and can be identified by tiny details: the spiral of the bellybutton nautilus, for instance, is visible at the center of the shell (hence, the scientific name *Nautilus macromphalus*, indicating a navel). And the fuzzy nautilus (*Allonautilus scrobiculatus*) has a shell that is partially covered by periostracum, a layer of thick fuzz.

Currently, nautiluses are the longest-living cephalopods, likely to survive for at least twenty years. Another unusual fact is that mature females reproduce once a year. Eggs resemble cloves of garlic, and juveniles hatch from them not as larvae, but as small but fully formed miniature copies of adults.

See page 90

EMPEROR NAUTILUS (*Nautilus pompilius*)

LIKE A RAM'S HORN

In the past, naturalists struggled to classify the spiral shells shaped like miniature rams' horns that they found on the beach. The shells had chambers inside but weren't large enough to house their owners.

Only later was it discovered that they belonged to the ram's horn squid (*Spirula spirula*), a species whose shape is similar to that of a squid. Because the ram's horn squid is a deep-ocean dweller, it has only rarely been observed by humans.

Its shell is very buoyant, which means it often floats to the surface after the mollusk has died.

Surprisingly, the ram's horn squid has its shell inside its body. Rather than serving as protection, the shell keeps it from sinking. Mystery solved!

See page 91

RAM'S HORN SQUID (*Spirula spirula*)

UNDERWATER PYJAMAS

On land we have zebras, but in the deep ocean we have the stylish black-and-white striped pyjama squid (*Sepioloidea lineolata*)!

This small cephalopod gets its name from its bizarre appearance—it has a white body with elegant, thin, dark, almost black lines running up and down its body. Though it is called a squid, it is actually closer to a cuttlefish.

The pyjama squid defends itself with a poisonous bite or by quickly changing color while expelling ink. And the pyjama squid has another ace up its sleeve—it can secrete toxic slime.

This helps it flee from many predators, especially the smaller ones.

See page 92

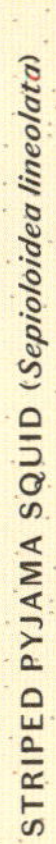

STRIPED PYJAMA SQUID (*Sepioloidea lineolata*)

AN UNDERSEA ORCHID

If we were forced to choose the most colorful cephalopod of all, it would have to be the flamboyant cuttlefish (*Metasepia pfefferi*).

Just a few inches long, it is white, black, yellow, and red, and it constantly changes color like a flickering campfire. Patches of color run from its head down its sides.

Looks can be deceiving, however: these bright colors appear only when the cuttlefish finds itself facing a predator or a scuba diver. Otherwise, it is a dull brown so that it won't stand out. Only when it is spotted does it signal its extremely poisonous bite to anyone even thinking about bothering it.

Its body and eyes are covered with small, fleshy lobes—lumps that blur the lines of its profile and make it less visible.

The flamboyant cuttlefish, sometimes described as an orchid because of its beauty, rarely leaves the sea floor and walks on its short arms.

See page 93

FLAMBOYANT CUTTLEFISH (*Metasepia pfefferi*)

AN EPIC GATHERING

Should you ever be in Australia between May and July, you might have the amazing opportunity to see cuttlefish congregating in the waters along the rocky coastline.

This is when Australian giant cuttlefish (*Sepia apama*), so named because they can grow up to 40 in. [1 m] in length including arms and tentacles, gather for mating.

The males, which are much bigger and more colorful than the females, challenge each other with chromatic messages while constantly changing colors, jetting around and chasing one another for short distances. Their goal is to protect the females, which are more reserved and smaller and often seek shelter in rocks or nooks and crannies in the seafloor.

One interesting fact is that some of the smaller males disguise themselves as females by changing color. Then they quickly dart past the larger males busy fighting over females and manage to mate undetected.

AUSTRALIAN GIANT CUTTLEFISH (*Sepia apama*)

MINIATURE JEWELS

While giant cephalopods loom large in the popular imagination, the truth is that the majority of species are small.

In fact, pygmy squids (*Idiosepius* sp.) are among the smallest cephalopods in the world. No longer than .5 to 1 in. [1.5 to 2.5 cm], each is smaller than a human fingernail.

Adorned with showy chromatophores that make them colorful and sparkly like jewelry, these tiny predators feed on very small shrimp. In turn, however, these diminutive hunters must try to pass unnoticed and to avoid being carried far from the ocean floor by the currents.

For this purpose, they have developed an amazing trick: they can literally glue themselves to algae thanks to a sticky substance produced by adhesive glands. The small squids of another genus (*Euprymna* sp.) also produce a sticky substance, then coat themselves in sand.

See page 95

PYGMY SQUID (*Idiosepius* sp.)

GIANT AND COLOSSAL

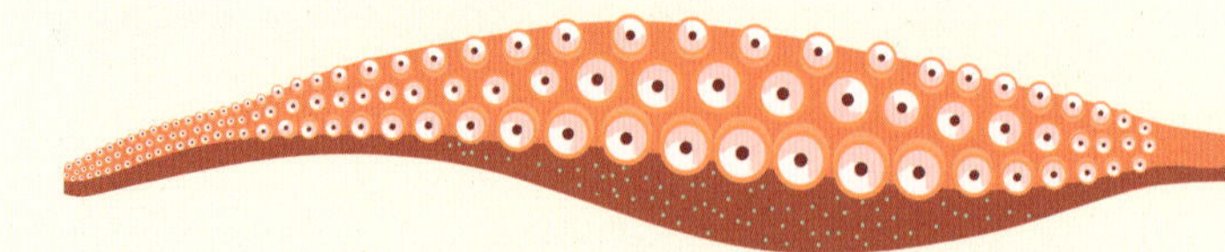

It's hard to imagine a marine animal more frightening than the giant squid (*Architeuthis dux*). This is a mollusk armed with two tentacles the length of a bus and eight arms that emerge from a body resembling a missile—not to mention two eyes as big as serving dishes.

Today we know that this animal is an active predator that lies in wait, living at least 1,000 ft. [300 m] below the surface and often even farther down. It feeds on other squids and fish. The largest examples are 50 to 60 ft. [15 to 18 m] long when all their tentacles are extended, and they weigh up to 1,100 lbs. [500 kg].

In 2007 an even larger squid was discovered—the colossal squid (*Mesonychoteuthis hamiltoni*). The colossal squid is even stockier and heavier (up to 1,550 lbs. [700 kg]), although shorter in length. It is found only in the deep waters of Antarctica, and it has hooks on its tentacles, as well as suckers—a detail that makes it truly frightening. It is the animal with the biggest eyes in the world (up to 16 in. [40 cm] in diameter). It is thought to use those massive eyes to identify prey and escape from its archenemy, the sperm whale.

In the last twenty years there have been some very exciting sightings of giant squids. In 2013, three scientists on a small submarine off the coast of Japan filmed a giant squid eating a diamondback squid in its natural environment 2,000 ft. [600 m] below sea level. Other fascinating encounters have taken place thanks to remotely operated vehicles that use luminous bait similar to jellyfish.

See page 96

LIKE A FEATHER

To study marine biodiversity, researchers catch species of every type so they can observe them up close and classify them.

Animals that live in the ocean depths are often injured as they come to the surface: an example of this is the chiroteuthid squid (*Planctoteuthis* sp.). For years this squid was described as having a classic squid shape.

Only recently, thanks to the use of remotely operated underwater vehicles, has it been possible to observe these animals alive and note something exceptional about them: the young especially are endowed with long, filiform, branchlike appendages at the rear that resemble feathers with spiral-like axes. This appendage looks something like a ship's propeller. It was hard to catch a glimpse of this previously because this part of the body easily breaks when the animal is caught.

No one knows whether the appendage has a specific function: maybe it helps the squid resemble a siphonophore, a gelatinous organism that stings, which would keep predators away.

See page 97

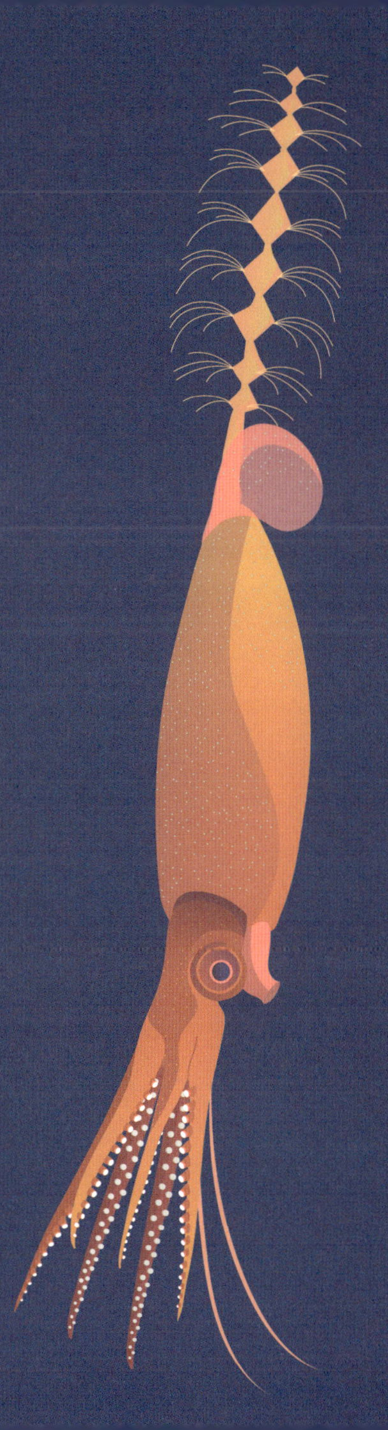

CHIROTEUTHID SQUID (*Planctoteuthis* sp.)

AN AQUATIC PIGLET

Sometimes we create associations between animal species in our own minds. But sometimes there really is a likeness!

The piglet squid (*Helicocranchia* sp.) is pink in color with prominent black eyes. It keeps its arms in an upward position as it glides through the water. The position shows off its siphon, which resembles a pig's snout. It shoots jets of water from the siphon, and the propulsion puts it in motion.

Not much is known about this animal's biology. It has been observed that the larvae live only a few hundred yards below the surface, moving farther down as they grow.

Males tend to have stronger arms and larger suckers than females.

See page 98

PIGLET SQUID (*Helicocranchia* sp.)

GLASS SCULPTURES

Sometimes, to go unnoticed, animals change color, adapting to the environment around them.

Others have landed on the clever solution of simply becoming transparent.

This is the strategy that evolution has developed for the glass squid, whose appearance is quite amazing.

The googly-eyed glass squid (*Teuthowenia pellucida*) is characterized by a light blue body that you can see right through because of its amazing transparency. Only its eyes and mouth are white and opaque. These squids are so transparent you can even see what they've eaten!

Glass squids use an ingenious trick to defend themselves: they swell so they look more round and harder to swallow.

If a predator doesn't give up, the squid can pull its head and arms back into its body and even fill up with ink. This makes it resemble a black ball in the unfathomable darkness of the deep ocean.

See page 99

GOOGLY-EYED GLASS SQUID (*Teuthowenia pellucida*)

TWO VERY DIFFERENT EYES

Many animals that live in the depths have big eyes so they can capture the small amount of light that is available and identify their prey even in the dark.

The strawberry squid (*Histioteuthis heteropsis*) is highly specialized: its right eye is small and light blue, while its left eye is twice as large, bulging, spherical, and yellow.

Why? So that it can see better!

This species needs a sophisticated instrument to probe the penumbra, so it uses its yellow eye (which has a natural filter) to eliminate the dominant blue of the depths and better identify the bioluminescence of its prey.

Its body, whose color and pattern recall that of a strawberry, has various luminous organs that probably serve to attract prey, making its work even easier. It also has a sophisticated system that regulates the intensity of its luminescence based on the surrounding environment.

See page 100

STRAWBERRY SQUID (*Histioteuthis heteropsis*)

UNIQUE AND RARE

Even given the great variety of shapes and colors that characterize cephalopods, Joubin's squid (*Joubiniteuthis portieri*) is definitely unique: not only are its arms long and thin but its body is filiform and slender, too, with small eyes and a mysterious branched appendix sticking out from its rear end.

Its red coloring helps it to camouflage against a dark background, since red begins to fade just a few yards below the surface and becomes totally invisible in the depths.

The rare images of this species obtained by researchers show these animals to be stationary in the middle of the water, their arms wide open as they wait for small prey, and also show that they move slowly even when they are disturbed.

See page 101

JOUBIN'S SQUID (*Joubiniteuthis portieri*)

HUGE FINS

When bigfin squids (*Magnapinna* sp.) were caught for the first time, in the twentieth century, they did not cause a stir: the individuals were injured, which made it hard to see just how beautiful the species was. Only later, thanks to videos produced from submarines and remotely operated vehicles, was it possible to see the true beauty of these cephalopods from the depths, which had been unknown until then.

The body of a bigfin squid is equipped with two very large round fins (as the name suggests), which allow it to soar like a butterfly above the seabed. Their tentacles and arms are thin and look very much alike. They hang down to catch small organisms. The bigfin squid is very long: the biggest one ever sighted was estimated to be around 25 ft. [8 m] in length.

One of the best films of such a squid was made in 2021 by researchers from the NOAA (National Oceanic and Atmospheric Administration) at 1.5 mi. [2.4 km] below sea level off the coast of Florida. In it, the squid appears to have a red body, with tentacles and arms that are red at the base while most of the filaments are white. In older videos the bigfin squid had appeared to be white.

As there have been only a dozen sightings in all, there is still plenty to find out about this elusive animal.

See description on page 102

BIGFIN SQUID (*Magnapinna* sp.)

FLYING ACROBATS

It is a well-known fact that flying fish with large pectoral fins can leap out of the water and glide to escape their predators. But surprisingly, some cephalopods can do the same.

When a Japanese flying squid (*Todarodes pacificus*) swimming close to the surface senses danger, it forcefully sprays water by jet propulsion from its siphon. This allows it to bound forward, glide for dozens of yards, and then fall back down far away (hopefully!) from those who would like to eat it for supper; its predators are tunas and sharks.

This amazing superpower is possible thanks to the wide fins, resembling lobes, that increase the animal's lift and therefore its ability to almost float in the air.

Other theories suggest that this is a normal way to get around, because squids move more quickly outside of the water than in it.

See page 103

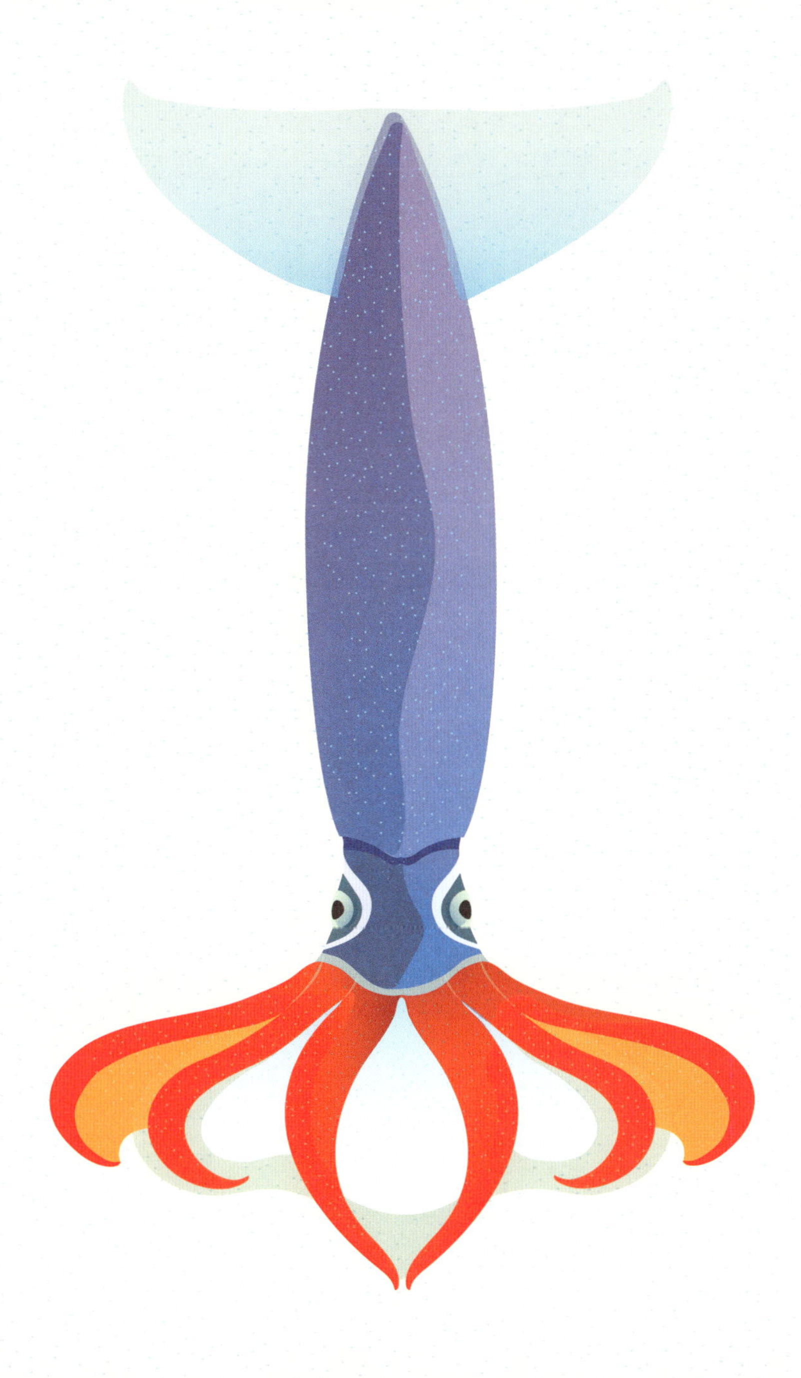

JAPANESE FLYING SQUID (*Todarodes pacificus*)

CURIOUS EGGS

The diamondback squid (*Thysanoteuthis rhombus*) can grow to considerable size and can weigh as much as 66 lbs. [30 kg]. It gets its name from its large fins, which make it look like a wide diamond; it is also known as the rhomboid squid.

Because there are not a lot of diamondback squids, they form monogamous couples even before reaching sexual maturity.

Males and females thus move together, and once they have reached adulthood, they spawn. The eggs are laid inside a protective slime structure that can be up to 40 in. [1 m] long!

Sometimes this mysterious object can be seen drifting on the surface of the water, arousing the curiosity of scuba divers and sailors. If you see one, take a picture of it!

These squids are often very elegant when young, with delicate webbing between their arms and a lively orange coloring.

See page 104

DIAMONDBACK SQUID (*Thysanoteuthis rhombus*)

AN INFERNAL VAMPIRE

Among the strangest species of cephalopods, and one of the most fascinating and famous, is the vampire squid (*Vampyroteuthis infernalis*). Bright red with triangular fins that it uses to get around, it has webbing that connects its eight arms, covered with soft protuberances resembling thorns (cirri) and suckers at the ends.

In spite of its scary name, it is harmless to human beings, as well as to most of the species that share its habitat. It feeds on suspended particles that it catches with its arms open like an umbrella. Even though it resembles an octopus, it has two tentacles like those of a squid, and it is, in fact, classified in a group of its own.

The vampire squid is capable of emitting light from the tips of its arms if attacked, and it can even release a bioluminescent cloud into the water. This is likely a defense strategy, aimed at distracting a predator's attention from the vital parts of its body. The luminous slime attaches itself to the attacker, making it more visible and vulnerable to other predators in the area, so that the hunter itself becomes the prey.

This species lives a few hundred yards below the surface, where there is a limited amount of oxygen. But thanks to its slow metabolism and large gills that help it to breathe, it manages to survive.

See page 105

VAMPIRE SQUID (*Vampyroteuthis infernalis*)

BABY ELEPHANTS AND GHOSTS

The invention of cable-controlled underwater craft and the use of small submarines for marine biology research have led to the discovery of new and virtually unknown worlds at the bottom of the sea.

One of the most singular characters on this stage in the dark depths is the flapjack octopus (*Opisthoteuthis californiana*) and its kin dumbos (*Grimpoteuthis* sp. and *Opisthoteuthis* sp.), so-called because of their large round fins, similar to the ears of the Disney elephant Dumbo.

Distinguishing among the species isn't easy, and some of them don't even have a name yet: this is because they've been observed only once, and only for a few minutes. But one thing is for sure: they've become famous since they appeared in the Disney Pixar movie *Finding Nemo*.

Researchers have unlimited imaginations, as does nature itself. One of the great mysteries yet to be fully explored is another creature named for a cartoon character, the Casper octopus, which is all white and resembles a ghost. Unlike the Dumbo octopus, it has no fins at all.

See page 106

FLAPJACK OCTOPUS (*Opisthoteuthis californiana*)

A TELESCOPE IN THE OCEAN

An astronomer uses a telescope to stargaze. Marine biologists have discovered an animal whose eyes remind them of that same instrument. The telescope octopus (*Amphitretus pelagicus*), widespread in the tropical and subtropical zones of the oceans, is relatively rare.

Its body is about 4 in. [10 cm] long, blue, and transparent, and its eyes are tubular—like telescopes—and red.

Little is known about this octopus, although scuba divers in Japan by chance spotted one close to the surface. Its calm behavior, expressed by its slow, almost pulsating movements, suggest that it imitates jellyfish or other gelatinous planktonic organisms like ctenophores.

See page 107

TELESCOPE OCTOPUS (*Amphitretus pelagicus*)

TRANSPARENT ARMS

The telescope octopus is not the only transparent pelagic octopod. Another is the glass octopus (*Vitreledonella richardi*). This species does not imitate organisms that sting but simply seeks to pass unnoticed: the only opaque part of its body (its digestive system) is cylindrical and is kept in a vertical position so that it is less visible from the bottom.

Its white suckers make it look rather distinguished.

Observation from submarines demonstrates that the male wraps its webbing around the female while mating. The female will then take the eggs everywhere with her because she can't attach them to the seabed.

Although the glass octopus was discovered in 1918, it is so hard to spot that researchers have managed to gather more information about it by studying its remains in the stomach of predators than by examining it directly.

See page 108

GLASS OCTOPUS (*Vitreledonella richardi*)

AN AUTHENTIC KRAKEN

People often speak about giant octopuses in general terms, but there's one specific sea creature that holds the record when it comes to size: the giant Pacific octopus (*Enteroctopus dofleini*). Though the seven-armed octopus (*Haliphron atlanticus*) gives it a run for its money, the giant Pacific octopus is on average the biggest octopus in the world: it can grow to a total of at least 10 ft. [3 m] and weigh over 400 lbs. [180 kg]. Some observers suggest it can grow even larger. The specimens observed are usually around 33 lbs. [15 kg], which is nothing to scoff at! Some researchers have reported seeing giant Pacific octopuses with suckers 2.5 in. [6.5 cm] across that can hold objects that weigh up to 35 lbs. [16 kg]. This allows them to capture unusual prey for cephalopods, like small sharks. For comparison's sake, the smallest octopus in the world, the star-sucker pygmy octopus (*Octopus wolfi*) is a mere 1 in. [25 mm] long.

See page 109

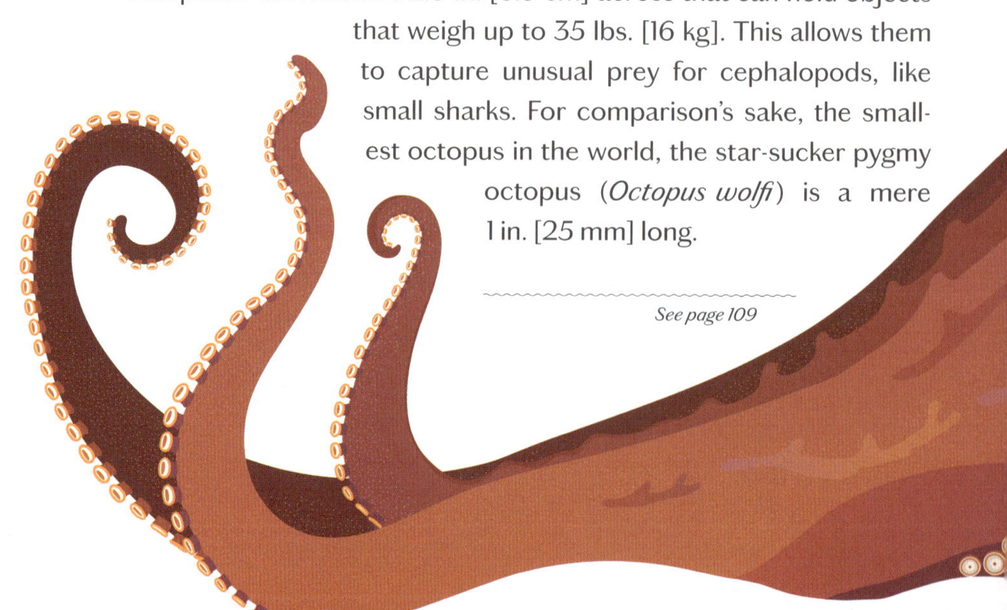

GIANT PACIFIC OCTOPUS (*Enteroctopus dofleini*)

A DOTING MOTHER

Although most octopuses have short lifespans, at least one species lives for a very long time: the warty octopus (*Graneledone boreopacifica*).

Observation at around 6,550 ft. [2,000 m] in depth has shown females caring for their eggs for four and a half years, which is much longer than the life cycle of all the other octopuses in the world. Most species die at somewhere between one and two years of age.

This octopus can live up to sixteen years, likely thanks to its slow metabolism, the cold water where it lives, and the reduced number of predators it encounters in the depths.

Its rose-colored mantle is covered in cartilaginous tubercles, which give the species its typically rough look, while its arms, which are often curled, have single rows of suckers.

Analysis of its diet has shown that it prefers gastropod mollusks. The muscles of its mouth are particularly well-developed so that it can break hard shells.

See page 110

WARTY OCTOPUS (*Graneledone boreopacifica*)

A HOUSE MADE OF COCONUT

The sandy seabed isn't a very good hiding place: if you're not flat or you have no way of hiding under the sandy substrate, you'll always be visible. That is why the coconut octopus (*Amphioctopus marginatus*) has developed a curious behavior—it looks for empty bivalves or split coconut shells and uses them for shelter.

When it finds only one half, it carries it around, holding it under its body like a skateboard and walking quickly on two arms; when it finally finds the other half, it uses its suckers to attach itself to the inner walls, closing itself off in a hard-shelled fortress. Clever!

The species is also known as the veined octopus because it is often a dark red color with black mesh that looks like veins. Its suckers are a stylish light blue.

See page 111

COCONUT OCTOPUS (*Amphioctopus marginatus*)

ELEGANCE IN RED

In the heart of the darkest night, the queen of octopuses moves along the seafloor. Known as the Atlantic white-spotted octopus (*Callistoctopus macropus*), this elegant species has incredibly long arms and is brick red. In case of danger, this color can become more intense and polka dots can appear all over. Its coloring as well as its ability to look bigger are part of what is known as deimatic behavior, aimed at intimidating its attackers. At the same time it makes this species of cephalopod one of the most beautiful in the world.

It lives in the Mediterranean and the Atlantic. With the help of an expert you might encounter one while swimming with a mask on and holding a flashlight. They may even be spotted close to the shore, where they hunt for small crustaceans. If you approach, they tend to scamper off, quickly slipping inside very small nooks and crannies.

See page 112

ATLANTIC WHITE-SPOTTED OCTOPUS (*Callistoctopus macropus*)

BEWARE OF THE BLUE RINGS

Walking along the beach in Australia you might come across signs that read BEWARE OF OCTOPUSES. Indeed, the notorious southern blue-ringed octopuses (*Hapalochlaena maculosa*), named for their beautiful colored rings against a yellowish background, live in tide pools. They are small, at most 4 in. [10 cm], including their arms. So, why should we beware of these stylish little animals?

Simple: their bites can be fatal for humans. Their saliva contains tetrodotoxin, a substance produced by bacteria. This quickly acts on the victim's nervous system and can kill an adult in two hours or less.

The southern blue-ringed octopus is not an aggressive species, but walking barefoot on the rocks or handling one could yield unpleasant results. Remember always to respect the inhabitants of the sea!

See page 113

SOUTHERN BLUE-RINGED OCTOPUS (*Hapalochlaena maculosa*)

THE YETI OF THE OCTOPUSES

You may be surprised to learn that many species of octopuses have not yet been fully described by scientists.

One of these is the hairy octopus (*Octopus* sp.), photographed and filmed by scuba divers in Indonesia and Japan. Just 2 in. [5 cm] long, it is distinguished from other species by long, branching papillae that covers its body and arms. The algae octopus (*Abdopus aculeatus*, widespread from Indonesia to the Philippines and Australia) resembles it, but their kinship is still awaiting confirmation.

No doubt its jaggedness contributes to its ability to camouflage. Those who have managed to glimpse this octopus say that if you lose sight of it, even for an instant, you've lost it forever—that's how hard it is to see, so keep your eyes open!

See page 114

HAIRY OCTOPUS (*Octopus* sp.)

NAMELESS BUT GREGARIOUS

Another species that is still to be classified is the harlequin octopus or the larger Pacific striped octopus (*Octopus* sp.), which still hasn't been given an official scientific name even though it has been known for decades.

Intricate and intriguing black patterns are visible against a white background. Its body is crossed by zebra stripes, but its arms feature leopard-like spots.

Although this may seem like just another cephalopod that can change color completely, becoming all dark or all light depending on the situation, its stripe-bar-spot pattern is unquestionably showy.

Unlike most octopuses, these mate by pressing suckers and beaks together in an intimate embrace. Nor is it a loner like other octopuses; it forms large social groups.

See page 115

LARGER PACIFIC STRIPED OCTOPUS (*Octopus* sp.)

THE TRANSFORMATION ARTIST

No animal in the world is more deceptive than the mimic octopus (*Thaumoctopus mimicus*), measuring around 20 in. [50 cm] from its head to the tip of its arms.

Because it is so beautiful, scuba divers and photographers actively search for it on the black sand of Indonesia, particularly in Sulawesi, where it is not as rare as elsewhere.

Much of its appeal lies in its amazing black and white stripes; they also make it very visible and easy for predators to spot.

Indeed, this pattern does not serve as a way to hide. Instead, the mimic octopus uses the pattern and its shapes and movements to imitate other striped species in the same habitat that are poisonous or toxic or bite.

Sea snakes, lionfish, and toxic soles are its favorite sources of inspiration. The mimic octopus manages to survive in seabeds lacking hideouts, where it can hunt crabs and mollusks right out in the open, even in the daytime.

See page 116

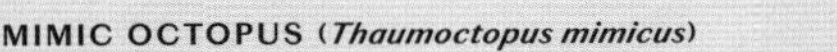

MIMIC OCTOPUS (*Thaumoctopus mimicus*)

THE WONDER OCTOPUS

While examining underwater photographs taken in Indonesia, researchers discovered a new species. Because it was rare, however, many years passed before it could be studied live.

The wunderpus (*Wunderpus photogenicus*, a compound of the German word *Wunder*, meaning wonder, and [*octo*]*pus*, referring to cephalopods) vaguely recalls the mimic octopus, but it is smaller, with webbing that stretches between its arms. Those arms are adorned with alternating reddish and white stripes, and it has a mantle featuring spots. So great is this animal's beauty that those who discovered it gave it the scientific name *photogenicus* because it always looks great in pictures!

Intriguingly, a comparison of the images shows that each of these octopuses has a different arrangement of spots and stripes on its body (like a fingerprint). This feature allows experts to verify the identity of a specific animal without catching or tagging it; all they need to do is take photos to study their movements.

Like other octopuses, the wunderpus is believed to imitate poisonous species to defend itself, even though it is a nocturnal animal. One was seen killing a mimic octopus in a lethal embrace while fighting over prey.

Like other species, if the wunderpus is in danger it can lose an arm, which explains why some individuals appear to be impaired or to have arms that have grown back but are thinner.

See page 117

WUNDERPUS (*Wunderpus photogenicus*)

AN OCTOPUS ODYSSEY

One very strange octopus does not live on the ocean floor like most of its relatives but instead constantly swims in blue waters on an unending odyssey. It is known as the greater argonaut (*Argonauta argo*).

Its skills as a swimmer allow it to move slowly, jet-propelled by the water coming out of its siphon. But its most amazing feature is its egg case: what might look like a shell is actually a container for eggs. It is white and very delicate, with a surface that looks carved, as if chiseled by a sculptor. Thin, tightly spaced ribbing makes it especially beautiful.

Vaguely similar in shape to an ammonite or nautilus shell, the egg case is embraced by the female with two arms and serves as a shelter for both the mother and her future offspring. However, it is of little use against its archenemy, the swordfish.

Measuring 12 in. [30 cm] across, the female is more than ten times the size of the male. The male uses a special arm that detaches from its body to mate with its companion.

Nighttime observation has shown argonauts attached to jellyfish: they eat a part of the animal's umbrella, but they also exploit the protection of its stinging tentacles.

See page 118

GREATER ARGONAUT (*Argonauta argo*)

FATAL ATTRACTION

Never touch what you're not familiar with because even the most harmless-looking animals can be full of surprises.

To survive in the middle of the ocean, where there is no chance of hiding under a rock or amid algae, the Indo-Pacific blanket octopus (*Tremoctopus gracilis*) has developed some remarkable tactics: young males and females carry the tentacles of the dangerous Portuguese man-of-war, a relative of the jellyfish whose sting is painful and can be fatal to humans.

For their part, when under threat, adult females try to ward off predators by waving the large colorful webbing that develops between their arms as if it were a warning flag. The blanket octopus has another strange feature: the female can grow to 6.5 ft. [2 m] in length, while the male doesn't grow to more than .75 in. [2 cm] at most! This means that the female becomes one hundred times bigger than the male, in order to produce more eggs. It is one of the most extreme examples of sexual dimorphism among all the animal species on our planet.

See page 119

INDO-PACIFIC BLANKET OCTOPUS (*Tremoctopus gracilis*)

CATALOG

COMMON NAME:

EMPEROR NAUTILUS

SCIENTIFIC NAME:
Nautilus pompilius

FAMILY:
Nautilidae

SIZE:
Maximum diameter of the shell 8.5 in. [22 cm]

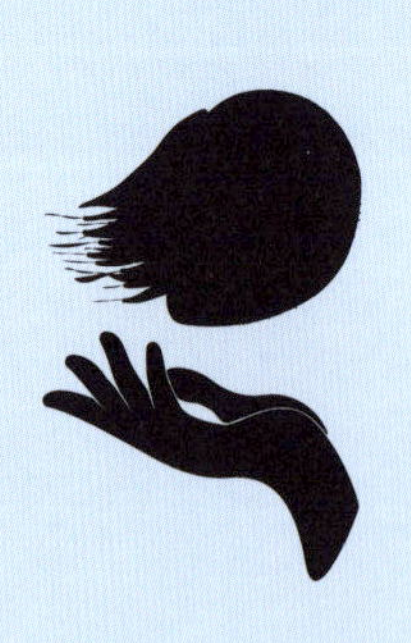

HABITAT:
Always rather close to the seafloor, in some areas at low depths but more often between 430 and 2,300 ft. [130 and 700 m]. The maximum depth reached is 2,625 ft. [800 m]—any lower and the shell implodes.

DISTRIBUTION:
Indian Ocean and western Pacific Ocean, from the Andaman Islands to Northern Australia, northward as far as the Philippines and Japan, although it is possible that the latter two areas are home to a similar species that has not yet been described.

FUN FACT:
Although it has reproduced in captivity since 1985, it wasn't until 2017 that the exact moment of a young emperor nautilus emerging from the egg was filmed in the Monterey Bay Aquarium in California. This has not yet been observed in nature.

See description on page 26

COMMON NAME:

RAM'S HORN SQUID

SCIENTIFIC NAME:
Spirula spirula

FAMILY:
Spirulidae

SIZE:
Body approximately 1.75 in. [4.5 cm] long

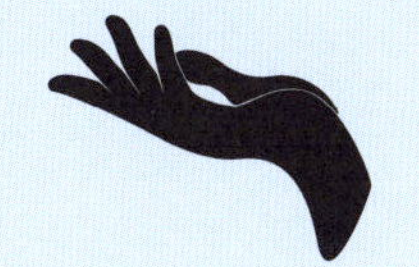

HABITAT:
Can reach depths of 1,640 to 3,275 ft. [500 to 1,000 m], rising at night. It can go as deep as 5,600 ft. [1,700 m], probably to reproduce closer to the seafloor.

DISTRIBUTION:
All tropical and subtropical waters

FUN FACT:
It was observed alive in its habitat only once (in 2020), in a vertical position with its head upturned.

See description on page 28

COMMON NAME:

STRIPED PYJAMA SQUID

SCIENTIFIC NAME:
Sepioloidea lineolata

FAMILY:
Sepiadariidae

SIZE:
Body around 2 in. [5 cm] long

HABITAT:
Sandy or muddy seabed

DISTRIBUTION:
Australia

FUN FACT:
Its pupils are located at the top of its eyes, probably to make it easier for it to see when it is buried in the substrate.

See description on page 30

COMMON NAME:

FLAMBOYANT CUTTLEFISH

SCIENTIFIC NAME:
Metasepia pfefferi

FAMILY:
Sepiidae

SIZE:
Body around 2.3 in. [6 cm] long

HABITAT:
Muddy seabed up to around 300 ft. [90 m] in depth

DISTRIBUTION:
Malaysia, Indonesia, Philippines, southward as far as Australia

FUN FACT:
Its white eggs become increasingly transparent as they develop.

See description on page 32

COMMON NAME:

AUSTRALIAN GIANT CUTTLEFISH

SCIENTIFIC NAME:
Sepia apama

FAMILY:
Sepiidae

SIZE:
Body around 20 in. [50 cm] long

HABITAT:
Rocky, sandy seabed, seagrass meadows, from the depths to over 325 ft. [100 m]

DISTRIBUTION:
Tropical and subtropical waters of Australia and Tasmania

FUN FACT:
The giant cuttlefish in Spencer Gulf seem to be genetically distinct, although currently they are not recognized as a species of their own.

See description on page 34

COMMON NAME:

PYGMY SQUID

SCIENTIFIC NAME:
Idiosepius sp.

FAMILY:
Idiosepiidae

SIZE:
Body less than
1 in. [2.5 cm] long for
most of the species

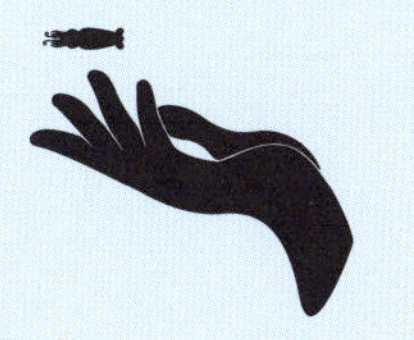

HABITAT:
Shallow coastal waters, especially those
surrounded by algae

DISTRIBUTION:
From the eastern coasts of South Africa to Australia,
northward as far as Japan

FUN FACT:
Currently only ten species are known to belong
to the pygmy squid group.

See description on page 36

COMMON NAME:

GIANT SQUID

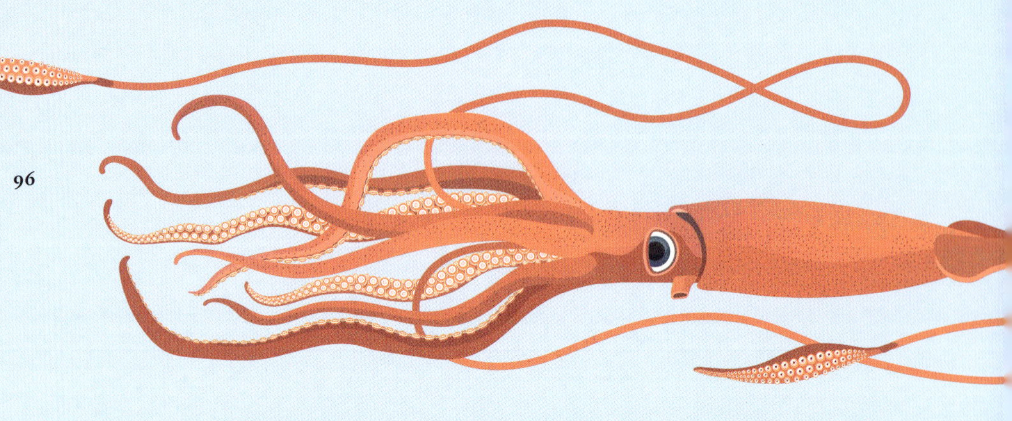

SCIENTIFIC NAME:
Architeuthis dux

FAMILY:
Architeuthidae

SIZE:
Body up to 6.5 ft. [2 m] long

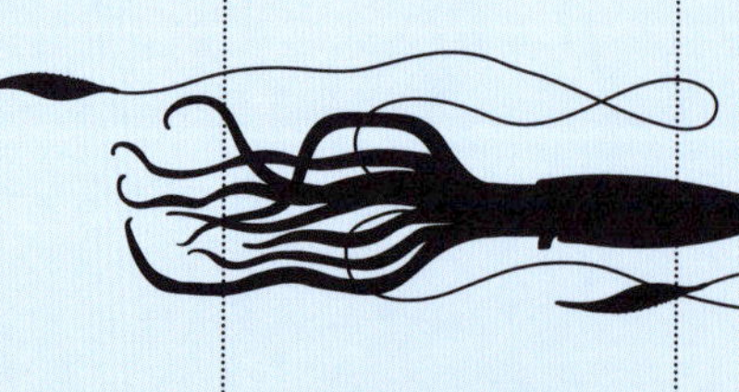

HABITAT:
Typically between 1,000 and 3,275 ft. [300 and 1,000 m] deep

DISTRIBUTION:
Oceans in temperate zones, especially the North Atlantic, South Africa, Australia, Japan, and the northeastern Pacific Ocean; a sighting reported in the Spanish Mediterranean

FUN FACT:
Although the giant squid is one of the largest animals on the planet, we know very little about its biology. In 2004 zoologist Tsunemi Kubodera and whale expert Kyoichi Mori were among the first in the world to observe one in its natural habitat, off the coast of Japan.

See description on page 38

COMMON NAME:

CHIROTEUTHID SQUID

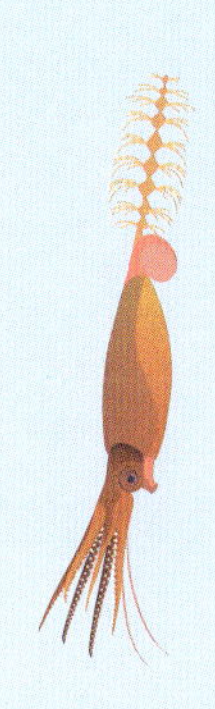

SCIENTIFIC NAME:
Planctoteuthis sp.

FAMILY:
Chiroteuthidae

SIZE:
Body between 1 and 3 in. [3 and 8 cm] long, depending on the species

HABITAT:
Generally from 2,625 to 7,875 ft. [800 to 2,400 m] below sea level

DISTRIBUTION:
All equatorial and tropical waters

FUN FACT:
Chiroteuthid squids are believed to be neotenic (i.e., continuing to resemble larvae even as they reach adulthood age).

See description on page 42

COMMON NAME:

PIGLET SQUID

SCIENTIFIC NAME:
Helicocranchia sp.

FAMILY:
Cranchiidae

SIZE:
Body 4 in. [10 cm] long in the largest species

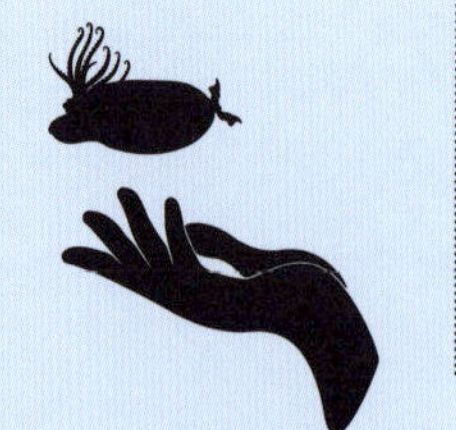

HABITAT:
Larvae closer to the surface, adults as much as 6,550 ft. [2,000 m] in depth

DISTRIBUTION:
All tropical and subtropical waters

FUN FACT:
Not all piglet squids are pink—transparent and bioluminescent specimens have also been observed.

See description on page 44

COMMON NAME:

GOOGLY-EYED GLASS SQUID

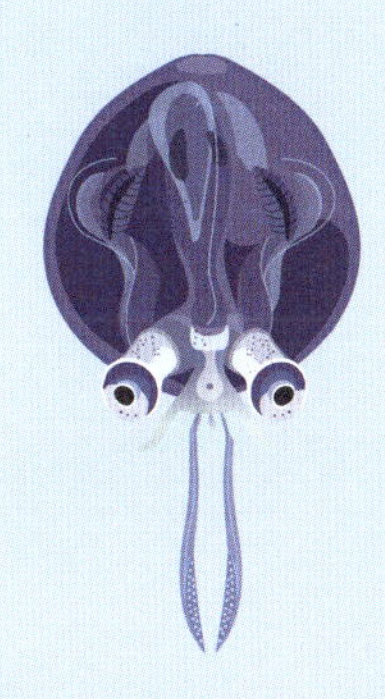

SCIENTIFIC NAME:
Teuthowenia pellucida

FAMILY:
Cranchiidae

SIZE:
Body around 8.25 in. [21 cm] long

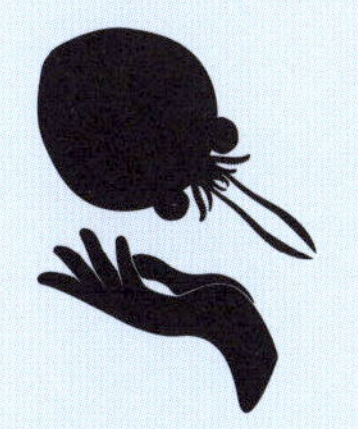

HABITAT:
Larvae at 650 to 1,000 ft. [200 to 300 m] below sea level. As they grow they migrate deeper; adults can be found at 5,000 to 8,850 ft. [1,500 to 2,700 m].

DISTRIBUTION:
Subtropical waters of Southern Hemisphere

FUN FACT:
It has been found in stomachs of hammerhead sharks, albatrosses, and sperm whales.

See description on page 46

COMMON NAME:

STRAWBERRY SQUID

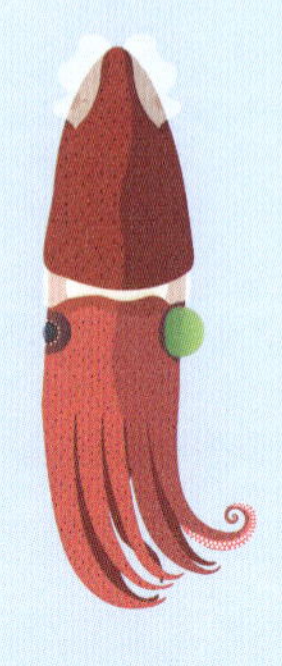

SCIENTIFIC NAME:
Histioteuthis heteropsis

FAMILY:
Histioteuthidae

SIZE:
Body up to 5 in.
[13 cm] long

HABITAT:
Between 1,000 and 2,625 ft. [300 and 800 m] below sea level, but moves up toward the surface at night

DISTRIBUTION:
Eastern Pacific Ocean

FUN FACT:
In both tropical and polar waters, numerous similar species have been spotted, all with asymmetrical eyes.

See description on page 48

COMMON NAME:

JOUBIN'S SQUID

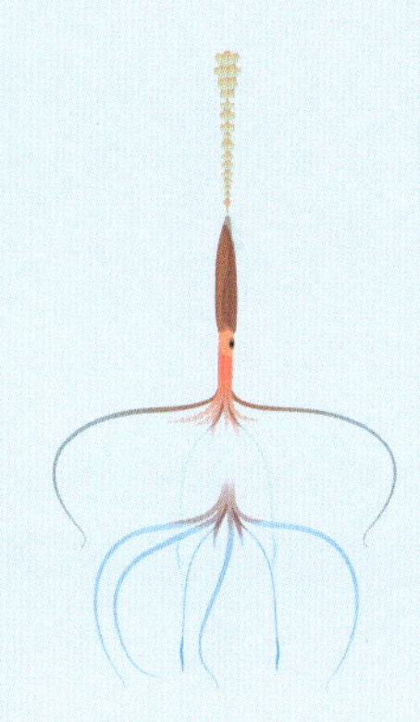

SCIENTIFIC NAME:
Joubiniteuthis portieri

FAMILY:
Joubiniteuthidae

SIZE:
Body around 4 in. [10 cm] long

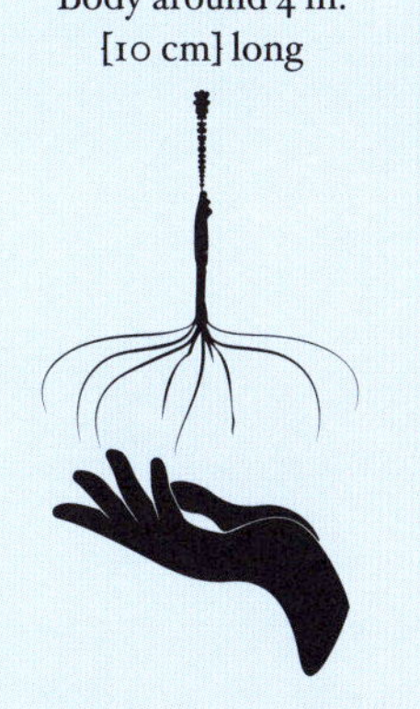

HABITAT:
Between 1,000 and 8,200 ft. [300 and 2,500 m] in depth

DISTRIBUTION:
All tropical and subtropical waters

FUN FACT:
This species is so unique that it has been classified in a family of its own.

See description on page 50

COMMON NAME:

BIGFIN SQUID

SCIENTIFIC NAME:
Magnapinna sp.

FAMILY:
Magnapinnidae

SIZE:
Body around 16 in. [40 cm] long

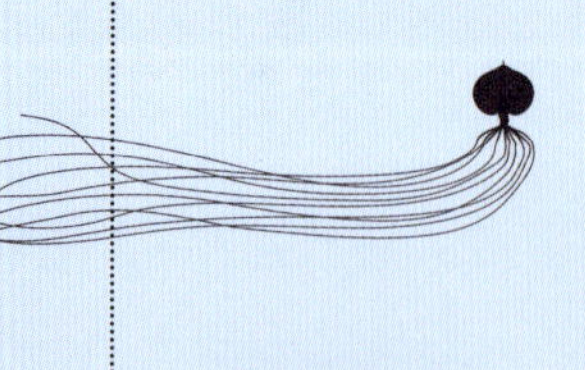

HABITAT:
The few individuals that have been sighted were between 6,365 and 15,530 ft. [1,940 and 4,735 m] in depth.

DISTRIBUTION:
Rare sightings of this animal in the Indian Ocean, the Pacific Ocean, and the Atlantic Ocean, but hardly anything is known about it.

FUN FACT:
One of the few young bigfin squids ever found was extracted from the stomach of a long-snouted lancetfish (*Alepisaurus ferox*).

See description on page 52

COMMON NAME:

JAPANESE FLYING SQUID

SCIENTIFIC NAME:
Todarodes pacificus

FAMILY:
Ommastrephidae

SIZE:
Body at most 20 in. [50 cm long], but usually shorter

HABITAT:
At around 650 ft. [200 m] deep in the ocean during the day, close to the surface at night

DISTRIBUTION:
Continental shelf from Japan to the western coast of North America, especially near deep seamounts

FUN FACT:
This migratory species follows ocean currents rich with nutritive substances.

See description on page 54

COMMON NAME:

DIAMONDBACK SQUID

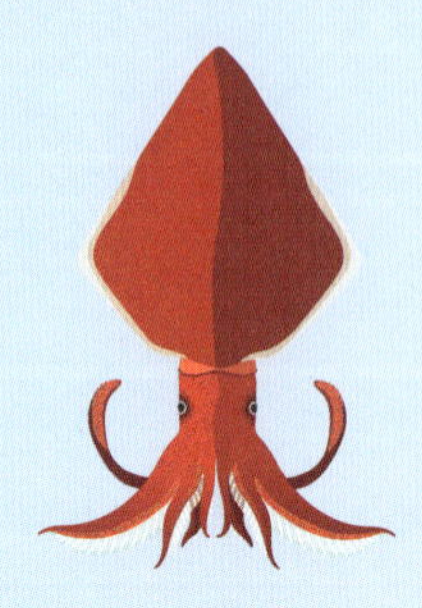

SCIENTIFIC NAME:
Thysanoteuthis rhombus

FAMILY:
Thysanoteuthidae

SIZE:
Body up to 4.25 ft. [1.3 m] long

HABITAT:
Offshore, where the seabed is at least 1,300 ft. [400 m] deep, and the surface water has a temperature of 68°F [20°C], mostly living between 1,475 and 2,130 ft. [450 and 650 m] in depth

DISTRIBUTION:
All tropical and subtropical waters, including the Mediterranean

FUN FACT:
The diamondback squid does not swim actively, but rather lets itself be carried by the current while steering with its fins, allowing it to save the energy it requires for its fast growth (gaining 37 lbs. [17 kg] in 300 days).

See description on page 56

COMMON NAME:

VAMPIRE SQUID

SCIENTIFIC NAME:
Vampyroteuthis infernalis

FAMILY:
Vampyroteuthidae

SIZE:
Body around 5 in. [13 cm] long

HABITAT:
Oxygen minimum zone, typically between 2,000 and 3,000 ft. [600 and 900 m] in depth, close to the seafloor

DISTRIBUTION:
All tropical and subtropical waters

FUN FACT:
In the nineteenth century, German biologist Carl Chun discovered the vampire squid during the Valdivia expedition. He had organized the expedition to prove that marine life existed 1,800 ft. [550 m] below the surface, thus dispelling the then-widespread theory that survival would be impossible at that depth.

See description on page 58

COMMON NAME:

FLAPJACK OCTOPUS

SCIENTIFIC NAME:
Opisthoteuthis californiana

FAMILY:
Opisthoteuthidae

SIZE:
Body around 3 in.
[8 cm] long

HABITAT:
Near the seafloor, from 400 to 4,000 ft.
[120 to 1,200 m] in depth

DISTRIBUTION:
Pacific Ocean, from Japan to the Bering Sea and along the American coastal waters as far as California

FUN FACT:
In some videos, members of this species appear to use their fins to hide or at least protect their bodies from potential threats.

See description on page 60

COMMON NAME:

TELESCOPE OCTOPUS

SCIENTIFIC NAME:
Amphitretus pelagicus

FAMILY:
Amphitretidae

SIZE:
Body around 4 in.
[10 cm] long

HABITAT:
Usually between 330 and 6,550 ft [100 and 2,000 m] in depth, far from the seafloor

DISTRIBUTION:
Tropical and subtropical waters of the Indian Ocean and the Pacific Ocean

FUN FACT:
The telescope octopus faces upward at all times to make itself less visible to predators below.

See description on page 62

COMMON NAME:

GLASS OCTOPUS

SCIENTIFIC NAME:
Vitreledonella richardi

FAMILY:
Amphitretidae

SIZE:
Body around 4.3 in. [11 cm] long

HABITAT:
Outside the continental shelf, always in deep waters, from the surface to at least 3,275 ft. [1,000 m] in depth

DISTRIBUTION:
All tropical and subtropical waters

FUN FACT:
A glass octopus larva has a sturdy beak that it uses to tear fragments of tissue off its prey, while the beak of an adult is less rigid because it eats softer prey.

See description on page 64

COMMON NAME:

GIANT PACIFIC OCTOPUS

SCIENTIFIC NAME:
Enteroctopus dofleini

FAMILY:
Enteroctopodidae

SIZE:
Body at least 25 in. [60 cm] long

HABITAT:
From tide pools to over 5,000 ft. [1,500 m] in depth, on rocky seabeds and, more rarely, sandy, muddy ones

DISTRIBUTION:
Northern coasts of the Pacific Ocean, from Japan to Mexico

FUN FACT:
The World Octopus Wrestling Championship was held in the 1960s. At these events, teams of scuba divers competed to catch the biggest octopuses. Today's divers are more respectful: they observe these fascinating mollusks in their habitat, but wisely keep their distance, since even the smallest, most innocent gesture—like a quick "embrace"—could lead to the loss of a mask or even a regulator!

See description on page 66

COMMON NAME:

WARTY OCTOPUS

SCIENTIFIC NAME:
Graneledone boreopacifica

FAMILY:
Megaleledonidae

SIZE:
Body around 5.5 in. [14 cm] long

HABITAT:
Rocky, muddy seabeds from 3,280 to 10,000 ft. [1,000 to 3,000 m] in depth

DISTRIBUTION:
Northern Pacific Ocean, from Japan to Alaska, southward as far as California

FUN FACT:
This is one of the most common cephalopods in its distribution area.

See description on page 68

COMMON NAME:

COCONUT OCTOPUS

SCIENTIFIC NAME:
Amphioctopus marginatus

FAMILY:
Octopodidae

SIZE:
Body around 4 in. [10 cm] long

HABITAT:
Can live at as much as 625 ft. [190 m] in depth, but it has mostly been sighted in shallow coastal waters and in muddy, sandy seabeds

DISTRIBUTION:
Coastal marine waters of the Indian and Pacific Oceans, from South Africa to the Red Sea and along the Asian coasts all the way to Japan and Australia

FUN FACT:
When threatened, a related species, the mototi octopus (*Amphioctopus mototi*), shows two large black spots with blue rings in the centers, known as fake eyes.

See description on page 70

COMMON NAME:

ATLANTIC WHITE-SPOTTED OCTOPUS

SCIENTIFIC NAME:
Callistoctopus macropus

FAMILY:
Octopodidae

SIZE:
Body around 6 in. [15 cm] long

HABITAT:
In shallow waters on both rocky and sandy seabeds and in seagrass meadows

DISTRIBUTION:
Mediterranean Sea and African Atlantic coasts as far as Senegal

FUN FACT:
A dozen similar species can be found in the tropical and temperate waters of the Atlantic Ocean, the Indian Ocean, and the Pacific Ocean. At present, very little is known about most of them.

See description on page 72

COMMON NAME:

SOUTHERN BLUE-RINGED OCTOPUS

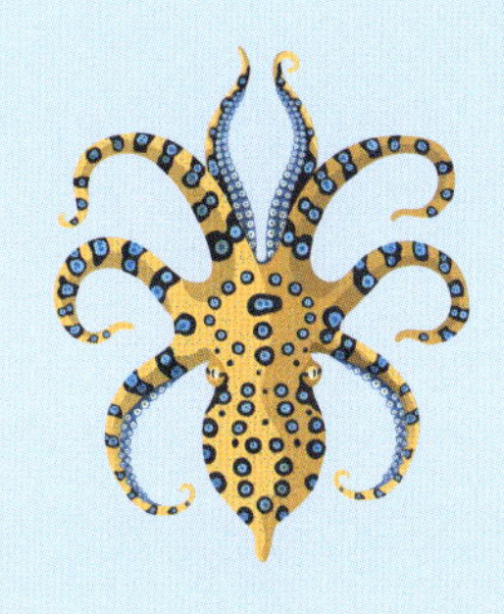

SCIENTIFIC NAME:
Hapalochlaena maculosa

FAMILY:
Octopodidae

SIZE:
Body around 2 in. [5 cm] long

HABITAT:
Coastal dweller that is commonly found from tidal pools to around 165 ft. [50 m] in depth, on rocky seabeds, coral reefs, and seagrass meadows

DISTRIBUTION:
Southern Australia and Tasmania

FUN FACT:
Females tend to mate with the largest males.

See description on page 74

COMMON NAME:

HAIRY OCTOPUS

SCIENTIFIC NAME:
Octopus sp.

FAMILY:
Octopodidae

SIZE:
Body around 2 in. [5 cm] long

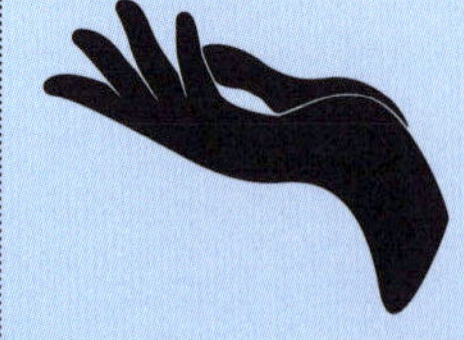

HABITAT:
Sandy seabeds with pebbles and shell fragments, around 65 ft. [20 m] in depth

DISTRIBUTION:
Sightings in Indonesia and Japan, and recently one in Madagascar

FUN FACT:
The area around the eye is often lighter in color, almost white.

See description on page 76

COMMON NAME:

LARGER PACIFIC STRIPED OCTOPUS

SCIENTIFIC NAME:
Octopus sp.

FAMILY:
Octopodidae

SIZE:
Body around 4.3 in. [11 cm] long

HABITAT:
Up to 1,000 ft. [300 m] in depth, but more frequently 130 to 165 ft. [40 to 50 m] in depth, on muddy and sandy seabeds

DISTRIBUTION:
Eastern Pacific Ocean along the Central American coasts, off Panama and Guatemala and in adjacent areas

FUN FACT:
One related species is the lesser Pacific striped octopus (*Octopus chierchiae*), whose males display behavior known as tasseling, i.e., quickly rotating the tips of their arms.

See description on page 78

COMMON NAME:

MIMIC OCTOPUS

SCIENTIFIC NAME:
Thaumoctopus mimicus

FAMILY:
Octopodidae

SIZE:
Body around 2.3 in. [6 cm] long

HABITAT:
Relatively shallow (130 ft. [40 m] or less) sandy and muddy seabeds, especially near river deltas

DISTRIBUTION:
Mainly Indonesia, but also Australia, New Caledonia, Papua New Guinea, Philippines, and to the west in the Red Sea and the Gulf of Oman

FUN FACT:
Unlike most octopuses, this species remains in its den all night long and resumes activity in the morning, hunting during the day.

See description on page 80

COMMON NAME:

WUNDERPUS

SCIENTIFIC NAME:
Wunderpus photogenicus

FAMILY:
Octopodidae

SIZE:
Body around 1.3 in. [3.5 cm] long

HABITAT:
Muddy seabeds no more than 65 ft. [20 m] in depth

DISTRIBUTION:
Indonesia, mainly Bali and Sulawesi, but sighted as far as the Philippines and Vanuatu

FUN FACT:
Like some other octopuses, the wunderpus uses the web-casting technique to trap its prey, spreading its arms wide like a spider's web to envelop it and then using its beak to bite it.

See description on page 82

COMMON NAME:

GREATER ARGONAUT

SCIENTIFIC NAME:
Argonauta argo

FAMILY:
Argonautidae

SIZE:
Female body around 4 in. [10 cm] long, egg case with a diameter of up to 12 in. [30 cm]; male body around .6 in. [1.5 cm] long

HABITAT:
Coastal marine waters, both near the surface and at depths of more than 1,000 ft. [300 m]

DISTRIBUTION:
All tropical and subtropical waters

FUN FACT:
For years, it was believed that the greater argonaut used its arms like sails on the surface of the water, stealing egg cases from other marine animals. French marine biologist Jeanne Villepreux-Power, who moved to Sicily in the nineteenth century, ferreted out the truth: this cephalopod secretes a substance from its arms that hardens and becomes a container for the eggs.

See description on page 84

COMMON NAME:

INDO-PACIFIC BLANKET OCTOPUS

SCIENTIFIC NAME:
Tremoctopus gracilis

FAMILY:
Tremoctopodidae

SIZE:
Female body around 12.5 in. [32 cm] long, male body around .6 in. [1.5 cm] long

♂ ♀

HABITAT:
Offshore, often near the surface, but has been spotted at up to 820 ft. [250 m] in depth

DISTRIBUTION:
Tropical and subtropical waters in the Indian Ocean and the Pacific Ocean

FUN FACT:
It is not easy to identify blanket octopuses because there are several similar species, one of which is *Tremoctopus violaceus*, which can also be found in the Mediterranean; to distinguish them one must carefully examine the pattern of the spots on the large stretches of webbing between the arms of adult females.

See description on page 86

TO LEARN MORE

BOOKS

Peter Boyle and Paul Rodhouse, *Cephalopods: Ecology and Fisheries*. Wiley-Blackwell, 2005.

Owen Davey, *Obsessive About Octopuses*. Flying Eye Books, 2020.

Stephanie Warren Drimmer, *Ink!: 100 Fun Facts About Octopuses, Squids, and More.* National Geographic Kids, 2019.

Craig Foster and Ross Frylinck, *Underwater Wild: My Octopus Teacher's Extraordinary World.* Mariner Books, 2021.

Peter Godfrey-Smith, *Other Minds: The Octopus and the Evolution of Intelligent Life*. William Collins, 2018.

Roger Hanlon, Mike Vecchione, and Louse Allcock, *Octopus, Squid & Cuttlefish: A Visual, Scientific Guide to the Oceans' Most Advanced Invertebrates*. University of Chicago Press, 2018.

P. Jereb and Clyde F. E. Roper (eds.), *Cephalopods of the World: An Annotated and Illustrated Catalogue of Cephalopod Species Known to Date.* FAO Species Catalogue for Fishery Purposes, 2005–2016.

Tsunemi Kubodera and Ryo Minemizu, *Cephalopods: Amazing and Beautiful Creatures.* X-Knowledge, 2014.

Sy Montgomery, *Secrets of the Octopus*. National Geographic, 2024.

Sy Montgomery, *The Soul of an Octopus: A Surprising Exploration into the Wonder of Consciousness*. Washington Square Press, 2016.

Mark Norman, *Cephalopods: A World Guide*. ConchBooks, 2000.

Claire Nouvian, *The Deep: The Extraordinary Creatures of the Abyss*. University of Chicago Press, 2007.

Danna Staaf, *The Lady and the Octopus: How Jeanne Villepreux-Power Invented Aquariums and Revolutionized Marine Biology*. Carolrhoda Books, 2022.

Danna Staaf, *The Lives of Octopuses and Their Relatives: A Natural History of Cephalopods*. Princeton University Press, 2023.

Danna Staaf, *Monarchs of the Sea: The Extraordinary 500-Million-Year History of Cephalopods*. Experiment, 2020.

Danna Staaf, *Squid Empire: The Rise and Fall of the Cephalopods*. ForeEdge, 2017.

WEBSITES

The Cephalopod Page
thecephalopodpage.org

Monterey Bay Aquarium Research Institute
mbari.org

OctoNation
octonation.com

Sea Change Project
seachangeproject.com

"Cephalopods," Smithsonian National Museum of Natural History
ocean.si.edu/ocean-life/invertebrates/cephalopods

TONMO
tonmo.com

FILMS AND VIDEOS

James Cameron and Steven Quale, dirs. *Aliens of the Deep*. 2005; Los Angeles, CA/Burbank, CA: Walden Media/Walt Disney Pictures, 2005.

Adam Geiger, dir. *Secrets of the Octopus*. Season 1, 3 episodes. Premiered April 21, 2024, on National Geographic Channel.

Natural World. Season 32, episode 2, "Giant Squid: Filming the Impossible." Aired July 13, 2013, on BBC.

Marcelo Johan Ogata, "Cephalopod Hallucination: Alien Intelligence," premiered on May 2, 2021. YouTube video, 49:01, 2021, https://www.youtube.com/watch?v=M9F66UsiJMc.

Anuschka Schofield, dir. *Animal*. Season 1, episode 4, "Octopus." Aired November 10, 2021, on Netflix.

Wildlife on One. Season 33, episode 4, "Gadgets Galore." Aired November 25, 2004, on BBC.

GLOSSARY

AMMONITES: Extinct class of marine cephalopods that mainly lived during the dinosaur age. These mollusks had spiral-shaped shells with intricate internal structures.

ARM: The arms of cephalopods have suckers used to grab and handle objects. The arms are often referred to as tentacles, but arms and tentacles are actually two distinct types of appendages.

BIOLUMINESCENCE: Ability of some organisms to emit light by way of chemical reactions in their bodies, commonly found in marine species like fish, jellyfish, and cephalopods.

CHROMATOPHORES: Specialized cells present in the skin of some animals, like cephalopods and chameleons. Chromatophores can expand or contract to change color, to camouflage the animal or to allow communication with other members of the same species.

CLASS: One of the hierarchical levels used in the taxonomy of living organisms. For animals there are seven main levels, from the most general to the most specific: kingdom, phylum, class, order, family, genus, species.

COLOR: Particular color pattern of an animal, often used to blend into the surrounding environment, attract partners, or indicate its emotional state.

CTENOPHORES: Gelatinous, transparent marine organisms of the phylum Ctenophora, with rows of cilia that they use to swim and feed themselves.

DECAPODIFORMES: Cephalopods with eight arms and two tentacles (for a total of ten appendages), like cuttlefish and squids.

DEIMATIC: Describes defensive behavior used by some animals to scare off or confuse predators. In cephalopods this often means spreading their arms wide, a sudden change in color, or the ballooning out of the body to seem bigger and more menacing.

EGG CASE: Structure produced by some animal species, like insects and mollusks, to contain eggs and offer protection during development.

GLADIUS: Internal flattened structure shaped like a sword made of chitinous material. The gladius supports the mantle when expanded and is typical of various squids.

HECTOCOTYLUS: The modified arm of the male of many cephalopods; it is used to transfer spermatophores to the female.

INK: Dark substance in cephalopods that is ejected as a defense mechanism to confuse or scare off predators.

KRAKEN: Legendary sea creature in Northern European tradition. Described as a huge squid or octopus, it is associated with sunken ships and storms at sea. It has also starred in some popular movies like *Pirates of the Caribbean*.

LIFT: Force that develops when an object, such as the wing of a bird or the fin of a marine animal, is immersed in a fluid, like air or water. It is what allows airplanes to take off and fly.

MANTLE: Organ consisting of elastic, muscular tissue that envelops the body of a cephalopod and contains its internal organs.

METABOLISM: The chemical reactions and biochemical processes that take place within an organism to maintain vital functions, such as the process of transforming food into energy and the use of this energy to support cellular activity.

MONOGAMOUS: Describes an animal species in which an individual has just one partner for a prolonged period or for its entire life.

NEOTENIC: Describes individuals that retain juvenile characteristics even in adulthood.

NEURONS: Specialized cells of the nervous system that are of essential importance for the transmission of information between the various parts of the body via electrochemical signals.

OCTOPOD: An eight-armed cephalopod, like an octopus.

PERIOSTRACUM: Outermost layer of the shell of many species of mollusks.

PHOTOPHORES: Specialized organs in some bioluminescent marine animals that allow them to emit light to attract prey, communicate with other members of the same species, or be less visible from below.

PLANKTON: Groups of small organisms that drift in the water, unable to propel themselves against a current. They are a key component of the food chain at sea. There are phytoplankton, with photosynthetic organisms like single-cell algae, and zooplankton, which includes small shellfish, the larvae of fish and mollusks, jellyfish, and the eggs of many marine organisms.

ROV (REMOTELY OPERATED VEHICLE): Underwater craft without a crew on board equipped with a camera and controlled remotely by an operator via a cable. ROVs are used in oceanic explorations, scientific research, and underwater work.

SEAGRASS: Marine flora that produces seeds and flowers, just like the ones on land. The most famous of these is *Posidonia oceanica*, which grows in a vast series of meadows in the water off the Mediterranean coast.

SIPHON: Tubular structure in the mantle of a cephalopod used for locomotion via the expulsion of water, for breathing, or for the release of ink in self-defense.

SIPHONOPHORES: Colonial marine organisms in the phylum Cnidaria, which includes jellyfish and corals, consisting of a colony of specialized octopuses. These include the famous Portuguese man-of-war (*Physalia physalis*).

TENTACLES: The two longest and most flexible appendages found in cuttlefish and squids, often with suckers, used to catch prey and to perceive the surrounding environment. In these animals, then, there are ten appendages (two tentacles and eight arms). Nautiluses instead show up to ninety tentacles.

TETRODOTOXIN: Highly potent poison of bacterial origin that acts upon the nervous system; it is present in the saliva of the blue-ringed octopuses and in the tissues of other marine animals, like the pufferfish.